The Vultures of Africa
A Coloring Book

Updated May 2016

Martha Nzisa Mutiso, Author
Keith L. Bildstein, Corinne Kendall, and **Munir Virani**, Editors
Wendy Frew, Graphic Designer and Illustrator

© 2016 Hawk Mountain Sanctuary Association

All right reserved. No part of this may be produced in any form, or by any electronic, mechanical, or other means, without permission in writing from the publisher, unless by non-profit groups, which may reproduce portions in non-electronic form, provided they credit Hawk Mountain Sanctuary Association.

Published by:
Hawk Mountain Sanctuary Association
1700 Hawk Mountain Road
Kempton, Pennsylvania 19529 USA
(610) 765-6961
www.hawkmountain.org
info@hawkmountain.org

Vultures of Africa

A Coloring Book

To learn more about African **vultures**, visit Hawk Mountain Sanctuary's website at:
 (1) www.hawkmountain.org. At Hawk Mountain Sanctuary's website you can go to "track migrants online" to see the movements of satellite-tracked Turkey Vultures and Black Vultures, and follow the blog "The Vulture Chronicles" to read about vultures and the people who are studying them.

Another website that has more information about African **vultures** is the Peregrine Fund's website at:
 (2) www.peregrinefund.org. At the Peregrine Fund's website you can go to the "Global Raptor Information Network" to learn more about the biology of vultures.

Other vulture websites that you might want to visit include the Endangered Wildlife Trust website at:
 (3) www.endangeredwildlifetrust.org where you can go to the "Birds of Prey Programme," and the Bird Life International website at:
 (4) www.birdlife.org where you can go to the "Data Zone" for information on the conservation status of vultures and other birds around the world.

(See the glossary at the end of this coloring book for explanations of words in **bold** type.)

Some books about vultures

Houston, D. (2001). Vultures & Condors. Worldlife Library Series, Granton-on-Sprey, Scotland.

Mundy, P. (1982). The comparative biology of South African Vultures. Vulture Study Group: Johannesburg, South America.

Mundy, P., D. Butchart, J. Ledger, and S. Piper. (1992). The Vultures of Africa Academic Press, New York, USA.

Sayre, A. P. Vulture view. (2007). Henry Holt and Company, New York, USA.

What are vultures?

Vultures are large, bald-headed **scavenging** birds that feed on **carrion**.

Vultures have hooked beaks that allow them to tear apart dead animals, and long broad wings that make them excellent flyers.

Vultures use winds and pockets of warm, rising air called **thermals** to soar long distances with little or no flapping.

Because the **carcasses** that vultures feed on are already dead, vultures do not have to catch and kill their prey. They do, however, need to find them.

Vultures have flatter feet and less curved talons than do other birds of prey that kill their food.

Vultures have featherless or nearly featherless heads and necks that are easier to clean after they have fed upon large carcasses.

Some benefits of vultures

Vultures help keep the environment clean by feeding on dead animals. In doing so they reduce the spread of many diseases.

Vultures are of cultural significance in many African communities.

Vultures often have high eco-tourism and bird-watching value.

Who are the African Vultures?

African Vultures belong to a group of **birds of prey** called "Old World vultures." Old World vultures are found in Asia and Europe as well as in Africa.

There are sixteen species of Old World vultures; eleven of which are found in Africa. The Cinereous Vulture and the Lappet-faced Vulture are the largest vultures in Africa.

The Cinereous Vulture is a North African **endemic** whereas the Lappet-faced Vulture is widely distributed within the continent.

The Hooded, White-headed, White-backed, and Palm-nut vultures are widely distributed in many African countries.

The relatively widely distributed Ruppell's Vulture, holds the altitude record for flying birds (approximately seven miles high), and travels long distances to look for food.

The Bearded Vulture, which has a limited distribution in Africa, is known for feeding mainly on bones and bone marrow.

The Cape Vulture is **endemic** to Southern Africa where its populations are decreasing. The Egyptian Vulture, which is found mainly in Northern and East Africa, is known for its ability to break and feed on eggs of larger birds.

The Griffon Vulture, which is widespread in Europe and parts of Asia, is found mainly in North Africa.

I am an adult Palm-nut Vulture
My scientific name is *Gypohierax angolensis*

What I look like

I am said to be "unmistakable" in the field, in that as an adult I am largely white except for black areas in my wings. I have a conspicuous red patch around my eye. In flight I am said to resemble an eagle more than a vulture.

Where I live

I breed in forests and savannahs, usually near water. I build my nest on palms, baobabs, and other tall trees. I lay one egg. I am quite approachable, and can be seen in and around human-dominated landscapes including hotel grounds and tourist areas in Gambia. I am found in thirty-six African countries including, Angola, Benin, Botswana, Burkina Faso, Burundi, Cameroon, Central African Republic, Chad, Congo, Côte d'Ivoire, Equatorial Guinea, Gabon, Gambia, Ghana, Guinea, Guinea-Bissau, Kenya, Lesotho, Liberia, Malawi, Mali, Mozambique, Namibia, Niger, Nigeria, Rwanda, Senegal, Sierra Leone, South Africa, Sudan, Tanzania, Togo, Uganda, Zambia, and Zimbabwe. South Africa boasts the largest population of Palm-nut Vultures in the world.

What I eat

I get my name from my primary source of food, which is not meat, but the nut of the Oil Palm tree. I also feed on fishes, crabs, amphibians, small mammals, and reptiles, as well as on carrion like most vultures.

Fun Fact

I am the only bird of prey that regularly eats nuts.

My Conservation Status

I am considered to be a species of **Least Concern**. I am widely distributed and relatively abundant in many areas.

Palm-Nut Vulture

I am an adult **Bearded Vulture**
My scientific name is *Gypaetus barbatus*

What I look like

I am a large vulture, and unlike most vultures, I have a feathered head. As an adult, I have a buff-yellow body and light-colored head, with a black eye-mask and moustache. My eye is yellow and I have a red eye ring. I sometimes rub iron-rich mud over my chin, breast and leg feathers, giving these areas a rust-coloured appearance. My feet are dark. My feathers are greyish black on my back, wings, and tail.

Where I live

I prefer rugged high mountains with numerous cliffs and gorges. I am found in eleven African countries including, Algeria, Egypt, Eritrea, Ethiopia, Kenya, Lesotho, Morocco, South Africa, Sudan, Tanzania and Uganda. I am occasionally reported in Djibouti, Mauritania, Mozambique, Namibia, Somalia, and Zimbabwe.

What I eat

Like other vultures, I scavenge food from the carcasses of dead animals. Unlike other vultures, however, most of what I eat is bones, including bone marrow.

Fun Fact

I am known to drop large bones while flying to break them into smaller, more eatable pieces. I also drop live tortoises to crack them open before eating them.

My Conservation Status

Although I am considered to be a species of **Near Threatened** species, my limited distribution and ongoing population declines makes me vulnerable as well.

Bearded Vulture

I am an adult **Egyptian Vulture**
My scientific name is *Neophron percnopterus*

What I look like

I am a small- to medium-sized vulture. As an adult my plumage is white, with some black feathers in the wings and tail. Because I often walk around large carcasses in dusty areas while waiting for my turn to feed on them, my bright white feathers are often dirty and brownish. My bare facial skin, which is yellow, turns orange during nesting periods. My tail is diamond-shaped, which makes me somewhat easy to identify in flight.

Where I live

I nest on cliffs and other rocky outcrops in lowlands and open country. I am found in thirty-three African countries including, Algeria, Angola, Benin, Burkina Faso, Cameroon, Chad, Côte d'Ivoire, Djibouti, Egypt, Eritrea, Ethiopia, Gambia, Ghana, Guinea, Guinea-Bissau, Kenya, Libya, Mali, Mauritania, Morocco, Namibia, Niger, Nigeria, Senegal, Somalia, South Africa, Sudan, Tanzania, Tunisia, Uganda, Western Sahara, and Zimbabwe. I am occasionally reported in Botswana, Congo, Lesotho, Mozambique, and Togo.

What I eat

I feed on animal carcasses, rotten fruits, and snails. On rare occasions I catch and eat live prey.

Fun Fact

I am a tool user. I break and eat the eggs of large birds like the Ostrich by picking up stones in my beak and throwing them at the eggs to crack them open.

My Conservation Status

I am currently considered to be **Endangered**. Major threats to my populations include unintentional poisoning, human disturbance, a decrease in food availability, overgrazing by livestock, power-line electrocution, and shooting at ostrich farms in South Africa.

Egyptian Vulture

9

I am an adult **Hooded Vulture**
My scientific name is *Necrosyrtes monachus*

What I look like

I am smaller than other African vultures. As an adult I have a mostly featherless, pink head and a greyish, overlapping hood. I have uniformly dark-brown body **plumage** with darker **flight feathers**. I have whitish patches on the sides of my **crop** and on my thighs. I have large, dark-brown eyes, surrounded by blueish eyelids. My ear opening is encircled by short hair-like feathers. My crop when filled and exposed is buffy with black feathers on its upper margin. My feet are greenish-white. I have broad wings for soaring and relatively short tail feathers.

Where I live

I am widely distributed and in West Africa. I nest on trees and lay one egg each year. I am found in thirty-eight African countries including, Angola, Benin, Botswana, Burkina Faso, Burundi, Cameroon, Central African Republic, Chad, The Democratic Republic of Congo, Côte d'Ivoire, Djibouti, Eritrea, Ethiopia, Gambia, Ghana, Guinea, Guinea-Bissau, Kenya, Liberia, Malawi, Mali, Mauritania, Mozambique, Namibia, Niger, Nigeria, Rwanda, Senegal, Sierra Leone, Somalia, South Africa, Sudan, Swaziland, Tanzania, Togo, Uganda, Zambia, and Zimbabwe.

What I eat

Decaying flesh from native animals and livestock is my main food. Sometimes I look for food in garbage dumps and marketplaces. I also feed on insects, including termites.

Fun Fact

I am a city bird in West Africa where I am common in towns and cities, but not in other parts of Africa, where I am more common in the field.

My Conservation Status

I have a wide distribution, and am currently listed as a **Critically Endangered** species. Recent concerns regarding declining populations in many parts of Africa may change all of that.

Hooded Vulture

I am an adult **White-backed Vulture**
My scientific name is *Gyps africanus*

What I look like

I am a medium-sized vulture. As an adult my head is a wonderful mix of colours. My face is generally pink, but can flush to red when I am excited, including when I am feeding on a carcass with other vultures. My **cere** and **gape** are pale powder blue, and my bill is bright orange. I have blackish eyes and legs. I have a white rump patch and white neck **ruff**. My plumage is brownish to cream colored.

Where I live

I am a lowland species of open wooded savanna, and particularly enjoy Acacia trees. I nest in tall trees. I am social and gather with other vultures at large carcasses, as well as when flying in **thermals** and at night-time **roosts**. I am found in thirty-seven African countries including, Angola, Benin, Botswana, Burkina Faso, Burundi, Cameroon, Central African Republic, Chad, Congo, Côte d'Ivoire, Eritrea, Ethiopia, Gambia, Ghana, Guinea, Guinea-Bissau, Kenya, Malawi, Mali, Mauritania, Mozambique, Namibia, Niger, Nigeria, Rwanda, Senegal, Sierra Leone, Somalia, South Africa, Sudan, Swaziland, Tanzania, Togo, Uganda, Zambia, and Zimbabwe. I am occasionally reported in Liberia.

What I eat

I eat the intestines and soft flesh of large dead mammals. I cannot tear thick skin but my long bill and neck, and my narrow head allow me to plunge deep into carcasses at openings and wounds.

Fun Fact

I am very social and often feed together with dozens of other White-backed Vultures.

My Conservation Status

I am considered to be a **Critically Endangered** species. My population has declined recently in parts of my range because of habitat loss, declines in populations of wild mammals, unintentional poisoning, and direct persecution. I also am hunted for trade in parts of Africa. In South Africa, for example, I am caught and consumed for perceived medicinal and psychological benefits.

White-backed Vulture

I am an adult Griffon Vulture
My scientific name is *Gyps fulvus*

What I look like

I am a large vulture with very broad wings and short tail feathers. As an adult my head and neck are covered in dense white down, except for a small naked area near the base of the neck. I have a pale yellow bill, lined with black along the cutting edges. I have a black cere. The skin of my face and neck is blueish, my eyes are a brownish yellow. I have a white neck ruff. My feet are black. My buff-colored body and wing **coverts** contrast with my dark flight feathers.

Where I live

I lay one egg per year or at a time, and nest on cliffs in small to large colonies. I am found in mountains in ten African countries, Algeria, Egypt, Eritrea, Ethiopia, Mali, Mauritania, Morocco, Senegal, Sudan, and Tunisia. I am occassionaly reported in Djibouti, Kenya, Libya, Niger, Togo, and Western Sahara.

What I eat

Like other vultures I am a scavenger, and feed mostly on dead animals that I find by soaring over open areas, often in flocks. I feed both outside of and inside the body cavities of large carcasses on the muscle, organs, and intestines of animals.

Fun Fact

I live in Europe and Asia as well as in Africa.

My Conservation Status

I am a species of **Least Concern**. I have a large range and population. I do however face problems, including land-use change and poisoning.

Griffon Vulture

I am an adult **Ruppell's Vulture**
My scientific name is *Gyps rueppellii*

What I look like

I am a relatively large vulture, and hold the record for the world's highest flying bird soaring nearly seven miles high. I have an overall grayish to blackish plumage with extensive pale creamy edging on my body feathers that makes me appear scaled or speckled. My wing and tail feathers are dark. I have a white neck **ruff**, greenish-grey neck, and pale, largely featherless head. I have a pale yellow eye and a strikingly yellowish-orange bill, with a small black tip. My **cere** is black.

Where I live

I am found in dry open country on cliffs and gorges, that are normally essential for nesting and roosting. I almost always nest on ledges in cliffs, and generally avoid human settlements. I am found in twenty-six African countries including, Algeria, Benin, Burkina Faso, Burundi, Cameroon, Central African Republic, Chad, Côte d'Ivoire, Djibouti, Ethiopia, Gambia, Ghana, Guinea, Guinea-Bissau, Kenya, Mali, Mauritania, Niger, Nigeria, Rwanda, Senegal, Somalia, Sudan, Tanzania, Togo, and Uganda. I am occasionally reported in the Congo, Egypt, Sierra Leone, and Zambia.

What I eat

I feed on the carcasses of dead animals. I locate my food by sight, either by scanning the ground directly when soaring above it, or by watching the activities of other vultures searching for food. I have an especially powerful bill and, after the most attractive soft parts of a carcass have been consumed, I sometimes continue to feed on the hide and even the bones.

Fun Fact

Recently I have been recorded breeding in Spain.

My Conservation Status

I am less common than previously believed, and am now considered to be **Critically Endangered**. Threats to my population include habitat loss through agricultural conversion, unintentional poisoning, nest disturbance, and direct persecution. I also am exploited for international trade in meat and medicine.

Ruppell's Vulture

I am an adult Cape Vulture
My scientific name is *Gyps coprotheres*

What I look like

I am a large vulture with a near-featherless head and neck. I am creamy-buff as an adult, with dark wing and tail feathers, and a pale buff neck-ruff. I have a yellowish eye, black bill, bluish throat and facial skin, and a dark neck. I am generally darker and more streaked when young, with brown to orange eyes, and a reddish neck.

Where I live

I nest on ledges on cliffs and usually lay one egg per year. I am found in five African countries including, Botswana, Lesotho, Mozambique, South Africa, and Zimbabwe. I also have been reported in Angola, the Democratic Republic of Congo, and Zambia. I have been **extirpated** from Namibia and Swaziland.

What I eat

I feed on carcasses of medium to large animals, mostly mammals from the size of Thompson gazelles (30-40kg) and larger.

Fun Fact

I breed farther south than any other African Vulture.

My Conservation Status

I am considered to be **Endangered** because I have a small population that appears to be declining. I am threatened by electrocution by power-lines, collisions with power-lines and vehicles, unintentional poisoning, food-stress during chick-rearing, direct persecution including killing for use in traditional medicines, and disturbance at **colonies**.

I am an adult **White-headed Vulture**
My scientific name is *Trigonoceps occipitalis*

What I look like

I am a medium-sized scavenger. As an adult I have a bright orange-red bill, pale blue **cere**, pale pink face and throat, and yellow eyes. I have a white **crest**. The featherless areas on my head are pale grey. My feet are bright pink. I have dark brown upper parts and black tail feathers. The feathers on my lower parts and legs are white.

Where I live

I build a stick nest in tall trees, and like to **roost** in tall trees near water. I lay one or two eggs that hatch after 56 days. I am found in open savanna and wooded country in thirty-seven African countries including, Angola, Benin, Botswana, Burkina Faso, Burundi, Cameroon, Central African Republic, Chad, Côte d'Ivoire, Congo, Djibouti, Eritrea, Ethiopia, Gabon, Gambia, Ghana, Guinea, Guinea Bissau, Kenya, Malawi, Mali, Mauritania, Mozambique, Namibia, Niger, Nigeria, Rwanda, Senegal, Somalia, South Africa, Sudan, Swaziland, Togo, Uganda, Tanzania, Zambia, and Zimbabwe.

What I eat

I lead a double life as both a scavenger and a **predator**. Sometimes I kill prey including small to medium-sized mammals, and locusts and other swarming insects. I also find and eat dead animals which I set out to find early in the morning.

Fun Fact

I am a generally solitary vulture that sometimes feeds on flamingo eggs.

My Conservation Status

As I am believed to be rarer than previously thought, and my conservation status was recently changed from **Vulnerable** to **Critically Endangered**. I face conservation concerns including decreasing populations of medium-sized mammals, habitat loss, indirect poisoning, human disturbance, and direct persecution. I also am killed for traditional medicine in South Africa.

White-headed Vulture

I am an adult Cinereous Vulture
My scientific name is *Aegypius monachus*

What I look like

I am the largest African Vulture. As an adult I have a large grey bill, with a blue cere and gape. My head is covered with dark-brown feathers. Parts of my head and neck are naked. My bare parts are bluish grey. My eye is golden brown. My feet are pale yellow. I have broad wings and a short, slightly wedge-shaped tail. Overall, I have a dark-brown plumage.

Where I live

I am found in forested areas in hills and mountains, and also in scrub and grasslands. I nest in trees or on rocks, often near other vultures in loose colonies. I am found in Sudan. I am occasionally reported in Egypt, Morocco, and Tunisia.

What I eat

I look for food over many kinds of open terrain, including semi-forested areas, in mountains, and in expansive grasslands. My diet consists mainly of the carcasses of medium-sized and large mammals. I occasionally prey on snakes and insects.

Fun Fact

I am more migratory than most people think. Individuals of my species that breed in Mongolia, for example, migrate all the way to Korea in the winter.

My Conservation status

I am considered to be **Near Threatened**. I have a small population which continues to decline due to both poisoned bait put out to kill dogs and other predators, and to higher hygiene standards that reduce the amount of carrion available.

I am an adult **Lappet-faced Vulture**
My scientific name is *Torgos tracheliotos*

What I look like

I am the second largest African Vulture. As an adult, my head is unfeathered and has fleshy folds, or lappets. I have dark eyes, a pale blue cere, a yellow bill, and red face. In flight, I am very dark with white thighs and a white bar running from my body to near the end of my wing.

Where I live

I build a huge, flat, grass-lined stick nest on tops of acacia trees. I lay one or two eggs which hatch after 54 to 56 days. I prefer undisturbed open country with some trees and little grass. I also am found on open-mountain slopes. I occur in thirty-two African countries including, Angola, Benin, Botswana, Burkina Faso, Cameroon, Central African Republic, Chad, Democratic Republic of Congo, Cote d'Ivoire, Djibouti, Egypt, Equatorial Guinea, Eritrea, Ethiopia, Gambia, Kenya, Malawi, Mali, Mauritania, Mozambique, Namibia, Niger, Rwanda, Senegal, Somalia, South Africa, Sudan, Swaziland, Tanzania, Uganda, and Zimbabwe. I am occasionally reported in Burundi, Libya, Morocco, and Togo. I have been **extirpated** in Algeria and Tunisia since the 1930s.

What I eat

I am often one of the first vultures to arrive at a large carcass. Smaller scavengers may rely on me to tear through the hide of a fresh carcass. In times of great need, I use my powerful beak to catch live prey, like young Thompson gazelles.

Fun Fact

I like to feed with my mate and often show up at a carcass in twos.

My Conservation Status

I am considered to be **Endangered**, I face conservation concerns including accidental poisoning through use of agricultural pesticides, nest disturbance, reduced food availability, habitat loss, and electrocution on power-lines. I often am mistakenly persecuted as a predator of livestock. I am sometimes hunted for food, medicine, and cultural reasons in West Africa.

Lappet-faced Vulture

Cape Vulture

Bearded Vulture

Griffon Vulture

White-headed Vulture

Egyptian Vulture

Lappet-faced Vulture

Palm-nut Vulture

Ruppell's Vulture

White-backed Vulture

Hooded Vulture

Cinereous Vulture

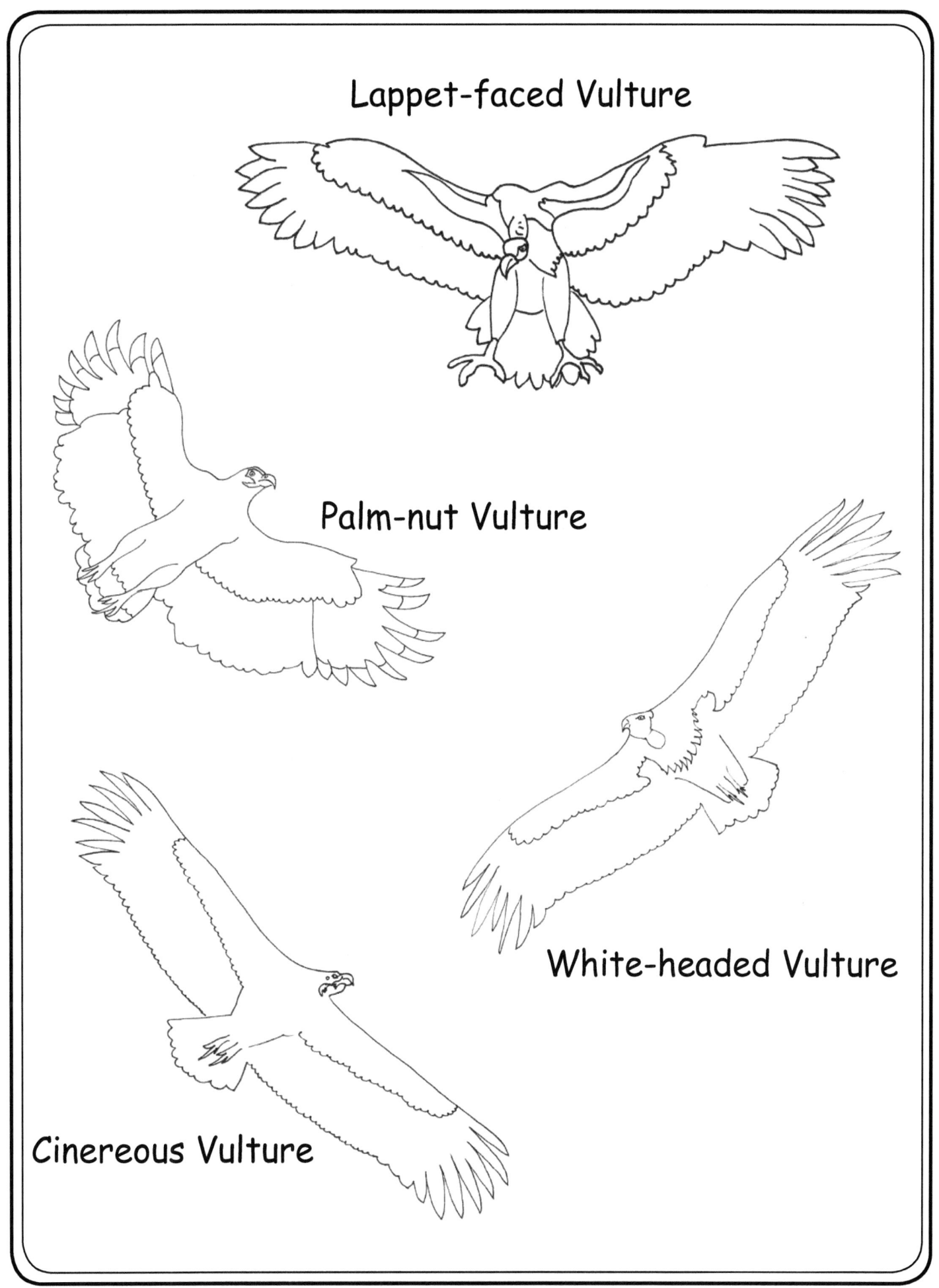

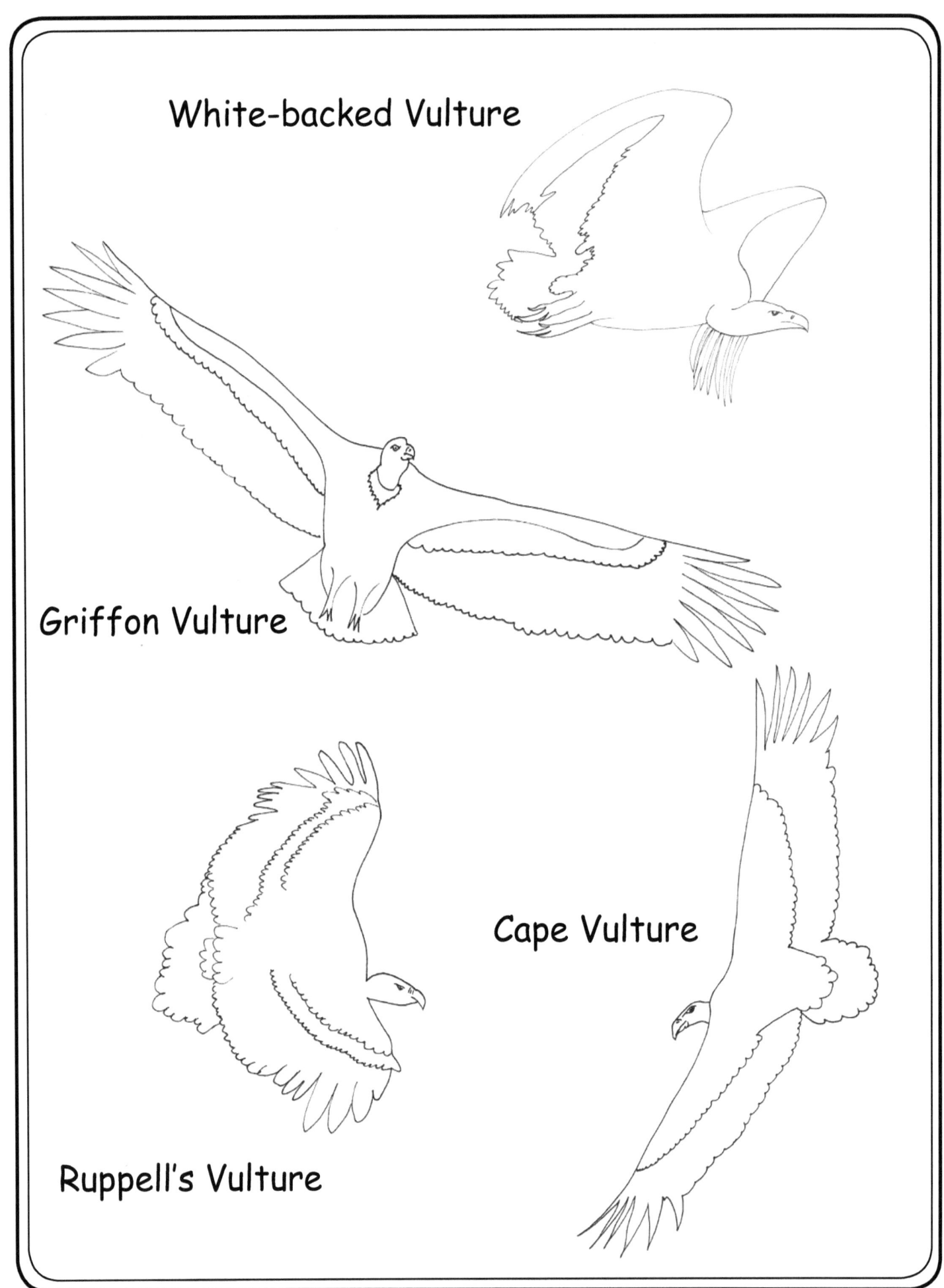

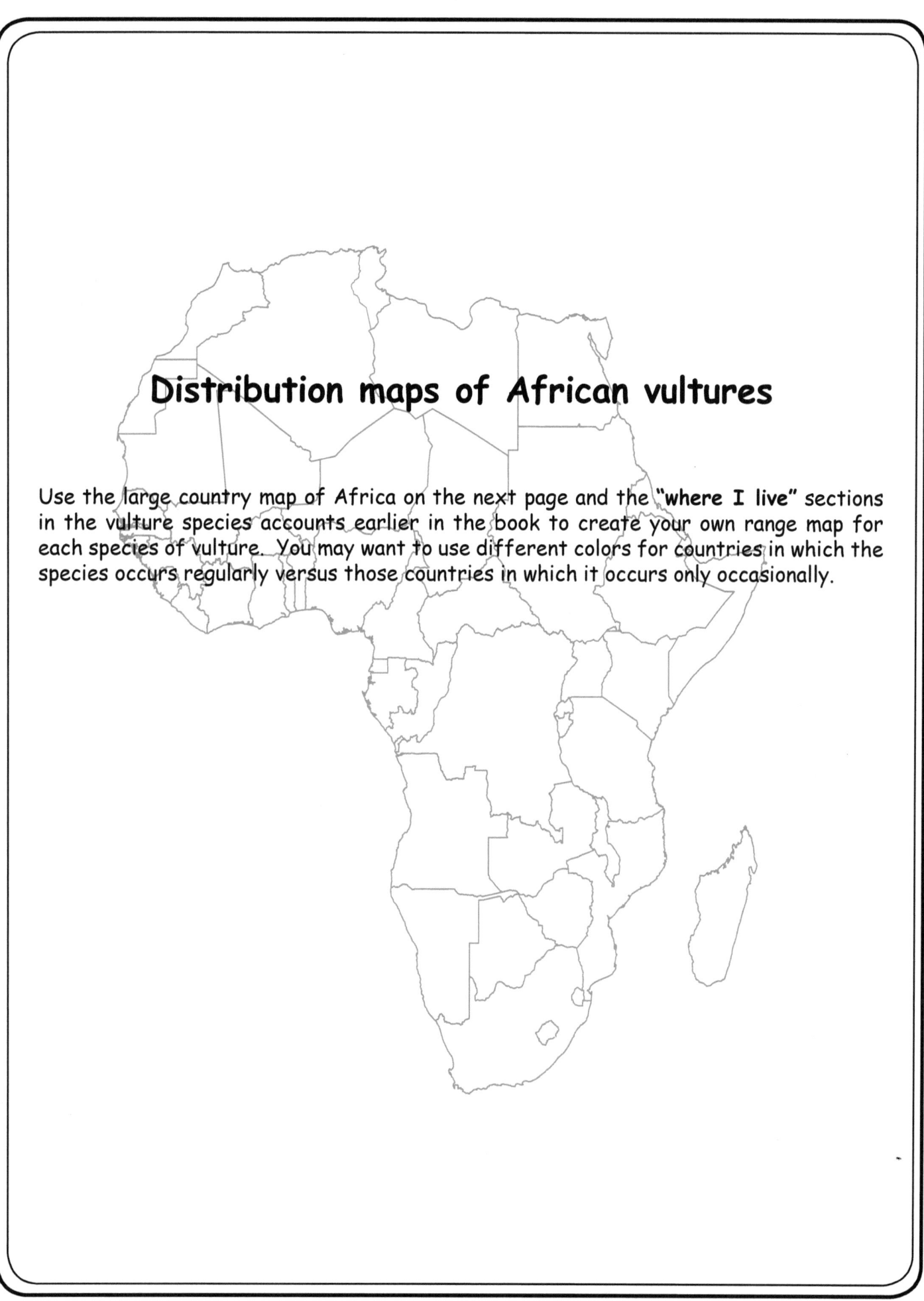

Distribution maps of African vultures

Use the large country map of Africa on the next page and the **"where I live"** sections in the vulture species accounts earlier in the book to create your own range map for each species of vulture. You may want to use different colors for countries in which the species occurs regularly versus those countries in which it occurs only occasionally.

MAP OF AFRICA

Palm-nut Vulture
range map

Bearded Vulture
range map

Egyptian Vulture
range map

Hooded Vulture
range map

White-backed Vulture
range map

Griffon Vulture
range map

Ruppell's Vulture
range map

Cape Vulture
range map

White-headed Vulture
range map

Cinereous Vulture
range map

Lappet-faced Vulture
range map

Glossary

Bird of prey: One of about 330 species of hawks, eagles, falcons, and vultures.

Carcass: The body of a dead animal.

Carrion: The flesh of a dead animal.

Cere: A soft, fleshy swelling surrounding the base of the bills of some birds, including vultures, through which the nostrils open.

Colony: A group of vultures that nests or roosts together.

Coverts: Feathers that cover the long flight feathers of the wing and that provide insulation and color.

Crest: A tuft of feathers on the peak of the head.

Crop: An expanded part of the upper digestive tract that stores food. The crop is often exposed as a featherless throat-bulge in vultures that have fed recently.

Endangered: A species that is facing a high risk of becoming extinct in the near future.

Endemic: Found only in a specific geographical area.

Extinct: A species that no longer exists.

Extirpated: A species that no longer exists in a portion of its historic range.

Flight feathers: The long feathers on the wings (i.e., primaries and secondaries) and tail (i.e., rectrices) of a bird.

Gape: Where the upper and lower jaws come together.

Habitat: The physical surrounding where a bird lives.

Least Concern: A species that does not qualify for Endangered, Vulnerable, or Near Threatened status.

Near Threatened: A species that is close to qualifying for Vulnerable status.

Plumage: A bird's feathers.

Predator: An animal that hunts and feeds upon other live animals.

Roost: A place where vultures perch to rest, usually at night.

Ruff: A projecting ring of down feathers around the neck of vultures that separates the largely unfeathered or featherless head and neck from the feathered body.

Scavenging bird: A bird that feeds on carrion, or the flesh of dead animals.

Soar: Flying without flapping wings while gaining alitiude or remaining aloft at the same altitude.

Thermal: A pocket or column of warm, rising air that birds, including vultures, can soar in. Rising pockets of warm air form when the sun heats different parts of the earth's surface at different rates. By soaring in thermals, birds save energy because they do not need to flap as much.

Threatened: A species that is at risk of becoming vulnerable.

Vulnerable: A species that is at high risk of becoming extinct in the mid-term future.

Vultures: One of 23 species of scavenging birds of prey.

African Vulture Checklist

Species	Where seen	When seen
Palm-nut Vulture		
Bearded Vulture		
Egyptian Vulture		
Hooded Vulture		
White-backed Vulture		
Griffon Vulture		
Ruppell's Vulture		
Cape Vulture		
White-headed Vulture		
Cinereous Vulture		
Lappet-faced Vulture		

Field Notes

www.ingramcontent.com/pod-product-compliance
Lightning Source LLC
Chambersburg PA
CBHW050835180526
45159CB00004B/1915